Memories of 1910

HALLEY'S COMET

ABBEVILLE PRESS · PUBLISHERS · NEW YORK

Memories of 1910

A HAPPY NEW-YEAR

by Roberta Etter
and Stuart Schneider

Editor: Walton Rawls
Designer: Jack Golden

Library of Congress Cataloging in Publication Data
Etter, Roberta B.
Halley's comet, memories of 1910.

1. Halley's comet—History. 2. Scrap-books.
I. Schneider, Stuart L.   II. Title.
QB723.H2E88   1985   523.6'4   85-15666
ISBN 0-89659-588-9

The authors wish to express sincere appreciation to Mr. Andreas Brown of the Gotham Book Mart for his support and encouragement on this project, toward which he has been more than generous in sharing his collection. The post cards that belong to him are: pages 28 (bottom), 36 (top), 53 (top), 58 (top), 59 (top), 62 (bottom), 69 (top), 72 (bottom), 73 (right), 75 (bottom right), 76 (top), 80 (far left), 82, 83 (top left), 90 (bottom), 91 (bottom), 95 (bottom left).

We are also grateful to Emily and Stewart Barr for their generosity in lending us items from their collection: pages 20 (top), 38 (top right), 50 (top right), 54 (bottom), 56 (right), 69 (bottom), 70 (bottom left), 84 (bottom right), 92 (top).

Copyright©1985 by Cross River Press, Ltd.
All rights reserved under International and Pan-American Copyright Conventions. No part of this book may be reproduced or utilized in any form or by any means, electronic or mechanical, including photocopying, recording, or by any information storage and retrieval system, without permission in writing from the Publisher. Inquiries should be addressed to Abbeville Press, Inc., 505 Park Avenue, New York, N.Y. 10022.
Printed and bound in Singapore by Tien Wah Press.
First edition.

A selection of rare and bejeweled Comet pins of 1910, all characterized by a "nucleus" and a "tail."

Table of Contents

Introduction	7
"Entire World Eagerly Waits Halley Comet"	19
"Experts in Doubt as to Effect of Comet on Earth"	31
"Hey! Look Out! The Comet's Tail is Coming Fast"	51
"On Lookout for Fiery Wonder; Tonight's the Night"	71
"Well, The Comet's Tale has Past, and Every One is Doing Business"	85

INTRODUCTION

It is easy to see how ancient man might be frightened by the sudden appearance of a comet. The skies above had always been predictable. Starry constellations remained in place in relation to each other, the sun shone brightly almost every day, and the moon went through its well-known cycles without faltering. Nevertheless, a fiery comet might appear without warning, and it was widely thought to be a genuine sign from heaven.

Indeed, comets were taken seriously all through history. The Greeks and Romans believed that the gods communicated with man through natural means, through the songs of birds, in the entrails of sacrificial animals, and by catastrophic natural events. A comet was seen as a forceful admonition from the gods.

In the year A.D. 79, Roman Emperor Titus Flavius Vespasian died beneath a "hairy" comet. Earlier, the comet of 43 B.C. was thought to be nothing less than the spirit of Julius Caesar on its way to the heavens, and the death of Agrippa, the Roman general, was marked by a comet in 11 B.C. The destruction of Jerusalem in A.D. 66 was foretold by a comet, as were the deaths of Emperor Constantine in 336 and Attila the Hun in 451. So great was the belief in comets as harbingers of a ruler's death that several chroniclers faithfully recorded the demise of Charlemagne as having been announced to the world by the comet of 814, a celestial event not attested to by others. In 837, a comet so frightened Emperor Louis I, who ruled both France and Germany, that he undertook a program of church and monastery building. In 1066, when a comet appeared before the Battle of Hastings, King Harold of England saw it as a sign of imminent defeat by the Normans, a phenomenon recorded in the famous Bayeux Tapestry.

On June 29, 1456, a comet sent people to their knees in deathly fear of plague, war, famine, and even the end of the world. As church bells pealed throughout the night, Pope Calixtus II ordered that the Hail Mary be recited three times daily and that the evil influence of the comet, a symbol of the "anger of God," be warded off by a short prayer, "Lord, save us from the Devil, the Turk, and the Comet!"

The English mathematician Thomas Digges, in summarizing common beliefs, was quoted in 1556 as saying, "Comets signify corruption of the stars. They are signs of earthquakes, wars, the changing of kingdoms, great dearth of corn, yea a common death of man and beast." In 1680, the French philosopher Pierre Bayle wrote that "The astrologer will tell you to what events the Comet has reference and the kinds of evils that may be expected. . . . An appearance in Aries signifies great wars; in Virgo, dangerous childbirth; in Scorpio, locusts; in Pisces, disputes concerning points of faith." He went on to say that "The astrologers will be right some day, and the public will remember that *one* prediction that has come true—better than all the rest that have proven false."

In 1680, Edmund Halley, who had been born in England in 1656, saw his first comet. In 1682, along with Sir Isaac Newton, he saw the comet that would later bear his name. He gathered data on the appearances and orbits of twenty-four comets and speculated that the comets of 1531, 1607, and 1682 were probably one and the same, orbiting the earth in a huge ellipse that brought it back about every seventy-five years. If it were so, Halley predicted that the comet of 1682 would return again in late 1758, and he so advised the Royal Society in London. Halley, who became Astronomer Royal in 1721, had urged Newton to write his great work *Principia* and even had financed its publication in 1687. Though Halley died in 1742, the scientific world remembered his prophecy and waited. In 1758, the comet returned as predicted and was thereafter known as Halley's Comet. Halley's greatest significance came from his deductions that comets were in orbit around the sun, but, unlike the circular orbits of the planets, comets traveled in great ellipses that took them thousands of millions of miles out into space. No longer would literate man wonder about the origin and meaning of comets; he could worry about the possibility that one of these fiery orbiters might crash into the earth.

A section of the 11th-century Bayeux Tapestry, showing the sudden appearance of a Comet, which King Harold of England took as a fatal omen (which it was!).

In 1773, a paper written by the eminent French mathematician Joseph de Lalande was misinterpreted to predict that a comet would hit the earth on May 20 or 21 of that same year. Reportedly, sales were made of seats in Paradise, to which certain of the clergy said they had obtained rights by special dispensation. However, doomsayers have flourished since the beginning of time, and predictions of the end of the world in a comet's catastrophic collision were made with regularity. *Weltuntergang*, as the Germans called it, was announced with great publicity for July 18, 1816, and then November 27, 1872, and then May 19, 1910—the fateful day featured in this collection.

Post cards of that era tended to mock the great fears of the world's coming to an end, but some of them might even have been serious, as in the case of the one showing a comet striking the earth while a dancing devil shouts a warning: "Make your Reckoning with the Heavens!" Whether viewed as good or evil, the comet has always lent itself to a tremendous outpouring of fantasy. On one card the comet was portrayed as a growling monster raging toward earth to consume its inhabitants; another card depicted the comet as a flaming Devil with huge talons who plucked innocent people off the earth as a sinister moon laughed in the background. On the other hand, a series of six colorful cards that came out both in French and German pictured a "smiling comet" approaching. As the comet drew nearer, the sun and moon were relieved that it missed them, but the reckless comet's path led it on into a collision with the frightened earth. The moon and sun were able to pull the comet's pointed head out of the earth, bandage it with a bright red and white polka-dotted kerchief, and send the tearful orbiter limping on its way back into limitless space.

The proliferation of illustrated post cards that greeted Halley's Comet on its last visit, seventy-five years ago, gives us a wonderful glimpse of early twentieth-century whimsy at its finest. Some of the cards detailed clever plans for escaping the earth's ultimate doom, including flights to the moon via the newfangled flying machine, or by rockets, balloons, floating inner tubes, and being shot from a cannon. The photographer on one card sticks up a large sign that announces his intention to offer rare snapshots of the catastrophe after the end of the "Spectacle." However, they will be available only through payment in advance.

The news clippings in this collection show that the papers were filled with "end of the world" items. Negroes were said to be gathering in the streets of Memphis, Tennessee, by the thousands, a prophet having told them that Halley's Comet would destroy the earth at high noon on May 19, 1910. In Kentucky, farmers neglected to plant their crops as they busily "got right with God" and made preparations for the end of the world. The Reverend Abraham Lincoln Johnson of Philadelphia "reduced his congregation to a paroxysm of fear" as he pictured for them the divine mission of the comet. Peasants in Italy were said to be terrified, and they attributed the unusual floods to the approach of Halley's Comet.

The appearance of Morehouse's Comet in 1908 had enabled scientists to analyze the chemical composition of a comet's tail, and they discovered the presence of cyanogen gas. They also calculated that in 1910 the earth would pass through Halley's tail as it swept by. Cyanogen combined with hydrogen produces prussic acid, and a single drop of that is enough to kill a person. With this news, the population of the world was in a real panic, and there were rumors that the tail also carried the influenza germ. How could people protect themselves from the poisonous gas? One leaflet bore the following advice:

Warning to the Inhabitants of the City
Close your windows and keep indoors
for the Earth will soon pass through
the Tail of the terrible Comet
and its poisonous gases will fill the Heavens!

Weltuntergang—the end of the world—was predicted to the exact day, and tremendous profits were to be made in exploiting this widespread fear of human extinction. Special "Comet Protecting Umbrellas," gas masks, "Anti-Comet Pills" (*Kometenpillen*), protective clothing, and excursions to the moon (payable in advance!) were marketed.

The comet was not always seen as a terrible menace by everyone; often it was viewed as a bringer of good luck and improved fortune. Comets were said to carry angels through the heavens or transport star-crossed lovers to more favorable climes. Many poems were written about Halley's Comet, including a particularly humorous one by Oliver Wendell Holmes (clipped for this collection) that appeared across the country in the newspapers of May, 1910.

The latest appearance of Halley's orbiter brought something entirely new: the Comet Party, which became the rage in 1910, both here and abroad. The society columns of the newspapers were filled with Comet Events, each designed to outshine the last. Stylish Parisians immediately saw the possibilities for festivities, and celestial costume parties became the latest thing. The invitation cards ran to the effect that "Monsieur and Madame request the honour of

A party hat created for the rash of Comet parties that took place in May, 1910, to celebrate Earth's passage through the tail of Halley's Comet.

(Right) A Royal Delft plate commemorating the year of Halley's Comet (Het Kometen Jaar). *Note the recognition of "1910A," an unexpected Comet that preceded Halley's that year.*

The comet could even be seen on the other side of the world!

your presence on the night of May 18 to mark the passage of Earth through the tail of Halley's Comet." Men were asked to wear pale blue evening dress and colorful scarves, while ladies were requested to don gowns "the color of the firmament and jewels to sparkle like the comet's tail."

Boston, New York, and other major cities of the United States also were alive with comet parties. All of the best hotels and restaurants joined in on "the social event of the season." The *Boston Globe* reported that "Drinks are being coined for the occasion, tables reserved, and extra decorations put on." Mr. Billy Smith of the Hotel Hayward made arrangements for six Comet Parties. Half of the tables were reserved well in advance, and the demand for more threatened to swamp the place. The Roof Garden of the Westminster was to be the scene of over a dozen individual parties.

Special decorations and party favors were crafted for the big event. The fancy hat in this collection features a smiling, giant sun flanked by two bright comets on the front, a crown of honeycombed paper, and crepe streamers that cascade out behind.

Among the drinks concocted for Comet Day consumption, the ingredients for the Comet Cocktail were kept a strict secret. However, it has been learned that basically it was a Rum Punch with a kick that would make the drinker see stars whether any were passing overhead or not! Halley's Highball was not guaranteed to refill its glass every seventy-five years, but its effect was promised to last for quite some time. The Nucleus Brandy Cocktail produced a sunbeam in every sip, while the Cyanogen Flip was guaranteed to do more damage than the Comet's tail and, perhaps, to cause as much merriment.

One other drink, that had a much longer association with the comet was champagne. To this day, certain well-known sparkling wines have a comet on their labels and even burnt into their corks. From the beginning of the nineteenth century, nature had

After a decade of very poor grape harvests in the Champagne district of France, a Comet suddenly appeared in 1811. The sparkling wines produced that year were regarded as spectacular, which led grateful vintners to add a Comet to their labels and corks in tribute.

STAR OF THE NORTH OR THE COMET OF 1861

The unexpected appearance of a Comet in 1861 was thought to have affected the course of the Civil War. This engraving appeared on a patriotic envelope supplied to the Union soldiers.

not been generous to the vineyards of Champagne. The vintage of 1805 had been judged "detestable," and the years 1808 and 1809 had not been much better. However, the year 1811 had produced an incomparable wine, both in quality and quantity. In this year a particularly impressive comet had appeared, and the superb vintage was baptized *Vin de la Comète*. Since that time, the comet has been used to identify fine champagne.

Throughout history, the appearance of a comet in the sky has also had a tremendous effect on politics. Suddenly, kings whose leadership had never been questioned might now be subject to the loss of their people's faith and allegiance. The comet permitted contenders for power to point skyward and declare the heavenly object to be a clear signal for the downfall of the reigning monarch. The observation, "He rode in on a candidate's coattails" is said to have had its origins in the expression, "He rode in on the comet's tail."

Abraham Lincoln was elected to the presidency of the United States at a time when a split between North and South was rapidly becoming a reality. Lincoln took the position that if the South seceded the U.S. government would fight to bring it back into the Union. The entire nation had its eyes upon Lincoln when a great comet appeared in the sky in 1861. It is said that the Southern leaders took it as a sign for action—a change in government. The comet was also seen by Mary Todd Lincoln, the president's superstitious wife. She is said to have greatly influenced the president's decision to go forward with his war plans. The Northern populace supported Lincoln's decision, and he was thought of as "The Star of the North" or "The Comet of 1861"; he was pictured as such on colorful patriotic envelopes supplied to the Federal troops.

In the year 1870, certain politicians focused on a new threat, which was graphically depicted as a comet by Thomas Nast in *Harper's Weekly* of August 6.

EDITORIAL PAGE OF THE
BOSTON AMERICAN

80 AND 82 SUMMER ST., BOSTON, MAY 18, 1910

The Flaming Comet

Few of Us Have Seen It, but All Are Interested, and Many Seem to Be Worrying About It. THE WORRY IS WASTED

Copyright, 1910, by American Journal Examiner

Flying around our sun and around our little family of planets is the great Halley comet. It drags behind it a tail millions of miles long, just as a flying locomotive drags a stream of smoke behind it.

This comet flies with frightful speed through the ether of infinite space.

Soon it will have disappeared. Nearly all of us will be dead before it comes back—about seventy-five years hence.

And during those seventy-five years it will continue rushing through the limitless space through which our globe is rolling.

We little human beings look out from this earth at the comet and we imagine many vain things. We are startled like the beetles, mice and toads in a cellar when the servant girl walks through with a lamp.

The thought in the mind of the beetle watching the lamp is probably about as intelligent as that in the mind of the average human being watching the comet.

One of our readers, Mr. J. J. Sanders of Prescott, Arizona, refuses to accept this newspaper's assurance that the comet won't hurt anybody. He writes: "I may be wrong, but I do not think so. I have warned our people to look out for earthquakes on the night of May 18."

Others believe that if the comet doesn't cause earthquakes it will cause diseases or fill our atmosphere with deadly gas and kill us all off.

None of these things will happen.

The comet will end its visit to our solar system as it has done hundreds of thousands of times before in the history of our young earth. Then it will go off, carrying with it its tail, twenty to forty million miles long, and we shall get over our excitement.

The earth MAY pass through the comet's tail and it may not. It won't matter, anyhow. If the comet happens to be near enough, and the tail of the comet happens to be long enough, this solid earth of ours will pass through the comet's tail to-night between 20 minutes past 11 P. M. and 20 minutes past 1 A. M.

The comet's tail is so nearly an absolute vacuum as to be scarcely worth thinking about. If you had a good vacuum pump and tried your best to produce a vacuum in an iron boiler, you would probably have, after you had finished, inside of that boiler more air per cubic yard than there is gas in the tail of that comet per cubic yard.

If you took a section of the comet's tail the size of Mechanics Building, and then took out of that section all the gas and all the solid matter it contained and fed it to a young baby, it wouldn't hurt the baby.

About one atom of solid matter or gas to the cubic yard is the only approach to solidity in that very diaphanous flaming tail.

Unfortunately, there are some superstitious, ignorant imaginations that cannot look upon anything unusual without fear. Their minds are rather close to the animal mind, and they are akin to the horse that shies a thousand times in succession at an automobile or a scrap of paper.

When our little earth rushes into the tail of the comet—if we do come in contact with the comet's tail—we shall simply brush against a few little particles of gas, and these will be instantly consumed by heat caused by contact with our atmosphere.

Remember that every year our earth is actually struck by something like forty millions of meteorites. We are bombarded every hour and every second by solid masses from the outside space, some of them huge rocks as big as houses.

Only once in a great while one of these strikes the earth and buries itself. Nearly always these solid rocks are melted when they strike our atmosphere and fall to the earth in harmless dust. Whenever at night you see a shooting star, so called, you simply see the flaming path of one of these meteors striking our atmosphere and melting as it falls.

How foolish for human beings, whose planet stands the incessant bombardment of solid rocks, to worry about the hollow tail of a gaseous comet because it happens to be unusual.

Get your mind off of the superstitious side of the comet discussion. Look at the interesting, inspiring and beautiful side of it.

That flying comet is a marvel worth thinking of, something to stir the mind and make us little earth insects proud to belong to this wonderful universe.

Actual facts are more interesting and more beautiful than foolish superstitions and fears.

The Halley comet is named for an Englishman named Halley. This man was first to predict the return of a comet.

About one hundred and fifty years ago he said to the other astronomers: "In a certain number of years from now, when I am dead, look out for this comet. It will appear at such a time, in such a place. If this prophecy comes true, remember that the prophecy was made by an Englishman."

The Englishman was perfectly right, the comet did return on time, as he had foretold.

Thousands of years ago the same comet was known and seen. It appeared 240 B. C. Robert Ball, the great astronomer, believed it to be the comet that appeared in 2616 B. C.

This great messenger flying through space, undoubtedly performing some useful function in this great universe (of which, as a whole, we know NOTHING), is marvelous in its grandeur and beauty.

The head of the comet is supposed to be nine thousand miles in diameter—exceeding by one thousand miles the diameter of our earth.

The comet's tail varies in length from twenty to forty millions of miles.

The speed is about twenty-five miles a second now, as it turns around the sun. It travels more slowly as it leaves the sun.

Some time we shall know what work this comet does, what part it plays in the life of the universe in which our sun and its planets are simply one little red corpuscle in the astronomical circulation of one corner of space.

Already the comet has been useful, even within our knowledge.

In 837 it frightened the Emperor Louis I., ruler of France and Germany. He built churches and monasteries, and behaved himself well.

It frightened the Saxons just before William the Conqueror landed in England, perhaps helped him to beat them. That was a good thing, for he took civilization to England.

Remember that this comet is really OUR comet. It's a sort of a "home body." It doesn't go very far away from us, only a little more than three thousand million miles.

There are other comets bigger and wilder that visit us for a little while and go off never to return. Some of these days, perhaps a million years from now, or day after to-morrow, one of these great flying bodies might possibly strike the earth and destroy us. But if it did, we should never know it, we should never have time to think about it, and, sad to relate, neither we little earth dwellers nor our solar system would be missed in the vast cosmic mechanism in which suns, planets, comets and nebulae play their part.

Brooch, buttons, bracelet, and compact, all displaying the Comet, as well as a box of Comet Blocks. The seal impression shows a bust of Edmund Halley with the Comet over his shoulder, and the pin with Robert E. Lee's portrait was made for a 1910 reunion of Confederate veterans.

The threat was the proliferation of cheap Chinese labor, pictured as a menacing, evil comet streaking across the sky and closely watched by Capitalists, the Press, Working Men, and Politicians. A large banner pleaded, "Trade Unions—Bull Against the Chinese Comet That Will Destroy Us!!!"

In 1910, politicians again chose to "ride the comet's tail," and that theme was used in a great number of campaigns. Former president Theodore Roosevelt was often pictured as the driving force of the 1908 and 1910 comets as he sought to return to a position of national power. Newspapers reported eagerly on the comet's effects on a candidate's future, and Governor Draper of Massachusetts was quoted as saying that he had "no more fear that the Comet or its tail will do harm than I have that the Democratic Party will be successful next fall!"

The advertising industry also found a partner in the comet. Since quite a few notable comets appeared in the nineteenth century they made news, and the right kind of news would catch the public's attention. If advertisers could get their products associated with a comet, hopefully all future comets would bring their companies' products to mind at purchase time. The most popular form of advertising in the last century was the trade card, and the most collectible of trade cards for that era used the comet theme. One of them advertised the Clipper Ship *Comet*, and the association was intended to imply extraordinary speed. By the 1880s, the trade card had advanced to a true art form, with bold graphics and vivid colors that competed for the public's eye. Since children as well as adults collected the cards and displayed them, merchants sought to produce the most eye-catching cards possible. Comets came to symbolize such diverse attributes as power, cleanliness, and strength, as well as speed.

Comets also appeared as incidental elements in the art of advertising. So much public attention was focused on the study of comets and their effects that

This Italian post card depicts the ancient belief that the appearance of a Comet is a sign from heaven that a despotic ruler should be overthrown.

One of numerous songsheets of the period associated with the Comet's appearance. A trip to the moon was a way of not being here if Halley's Comet was to collide with Earth. Another song was called "Halley's Comet Rag."

the mere image of a comet caught the eye. A particularly fine specimen of trade card from the 1880s shows a family gathered around a telescope on their roof. The verse at bottom reads:

> "What drives the Comet through the skies?"
> Inquiring little Johnnie cries.
> "Hold to the light my son and see—
> The Wonders of Astronomy!"

When the card is held to the light, the white area depicting the comet reveals a hidden drawing of "The Light-Running Jackson," a horse-drawn wagon manufactured by Austin, Tomlinson & Webster Mfg.

The brightness of the comet was used to symbolize cleanliness in a series of magazine ads for the Sapolio Soap Company that appeared in 1910. The theme argued that although Halley's Comet brightened the earth but once every seventy-five years, Sapolio (depicted as a bar of soap with a comet's tail) brightened the earth—every day!

There were Comet Brand matches, Comet Cut Plug Tobacco, Comet Fireworks, and the still very popular Comet Cleanser. Advertisements showed children studying comets through telescopes made of giant boxes of yeast or other products, and a comet streaked through the night sky as Santa Claus stuffed a huge spool of Clarke's Thread down a chimney. Earth was "raised" out of the path of a comet by the natural action of a large can of baking soda.

It is nice to think of Halley's Comet as a regular visitor to earth, for it marks notable steps in the progress of our civilization. When it appeared as late as 1758, we were just beginning to understand electricity through the experiments of Benjamin Franklin. In 1835, the railroad was in its infancy. In 1910, we were learning what it was like to fly. Who knows what our descendants will choose to characterize this visit by or what the world will be like—if a comet does not hit it!—at the time of Halley's next appearance?

ENTIRE WORLD EAGERLY WAITS HALLEY COMET

As Boston, New York and other cities of the world planned to-day to make to-morrow evening and the week-end a comet gala occasion, it was announced at Harvard Observatory that the tail of the celestial visitor is lengthening and spreading in fan shape, which insures that the earth will pass through the tail about 5,000,000 miles from its extreme end.

Professor David Todd of Amherst to-day said that the earth may be in the comets tail now. Most of the astronomers figure, however, that the earth will begin to enter the tail at 11:20 to-morrow night.

ALL SAY THE EARTH ISN'T IN DANGER.

All the students of the heavens agree that there will be no danger to the earth from the comet. Merry comet parties are being arranged all over the world. The people have manifested little if any fear of the daming visitor and its further appearance will be utilized as an opportunity for astronomical study and for festive gatherings.

On Thursday, Friday and Saturday evenings Boston may be able to get a good view of the comet shortly after sunset in the Western sky over the point where the sun went down.

At Harvard few of the astronomers remained up last night, because they did not expect to see anything of the comet. Those who did remain on watch had nothing to

Continued on Page 2, Column 3.

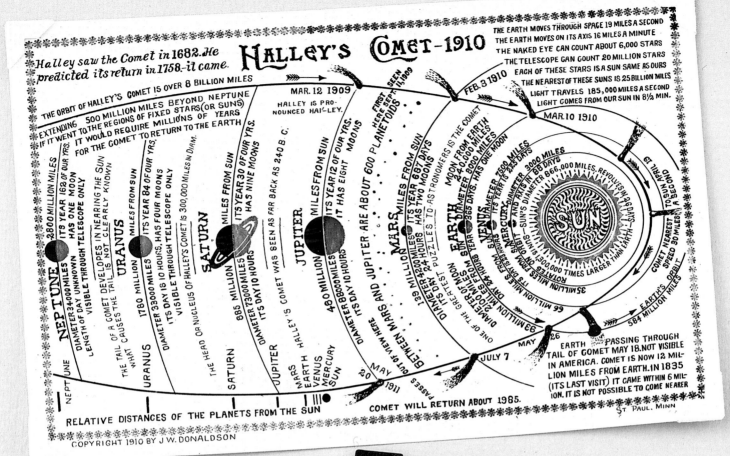

Most of us knew the comet was coming, but few knew where to look —

Béotisme Parisien.

La Comète de 1835.

Je vous assure, ma chère amie, qu'à ma Campagne de Bellevue, on voit bien mieux la Comète qu'ici.
— Je le crois sans peine, ma tante, Bellevue est sur une hauteur.

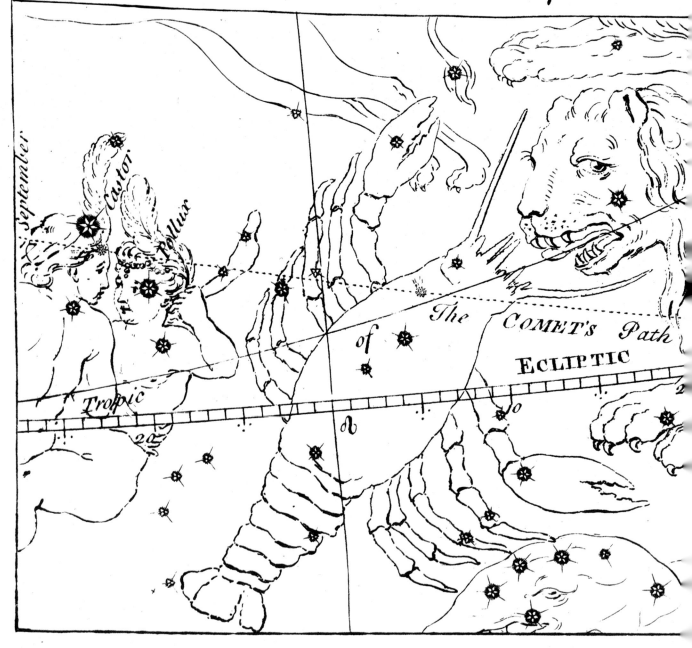

Human destiny was ruled and even forecast

...nt Year *1757* in its Return to the SUN.

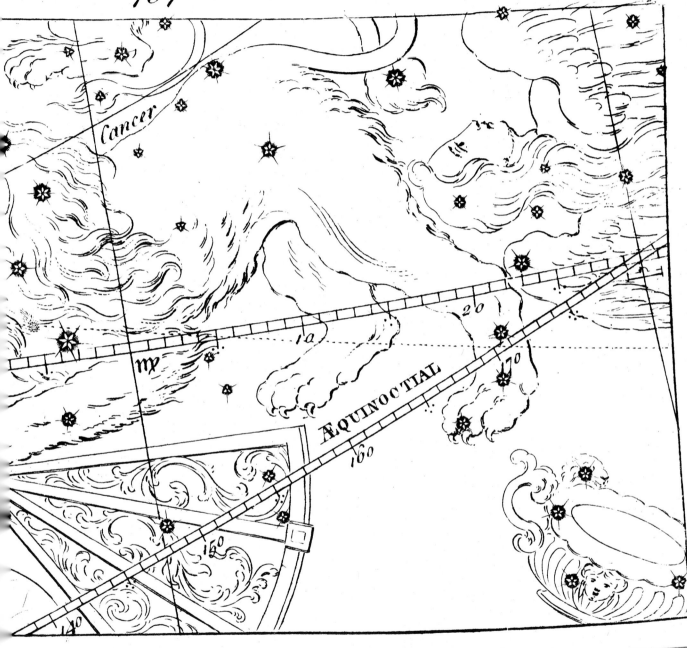

...events in the heavens.

DISCOVERY OF THE COMET—COME AT LAST! TEUTONIC ASTRONOMICAL PROFESSOR—"Mein Gott! it is—it is! Und its tail's black and waggles! Gott in hemmel, vot a Gomets!!"

HALEY'S COMET

Lecture on Popular Astronomy by
PROF. D. W. MOREHOUSE of Drake University
At First Baptist Church, Eighth and High Sts.

Tuesday Evening, May 17, 1910
FIRST DIVISION
15 Cents.

When Earth passed through the tail of the Comet in 1665 that was thought to have brought on the Black Death Plague.

CHANGES REPOR[TED]
COM[ET]

- GREAT COMET OF 1811. TAIL 100,000,000 MILES LONG. VISIBLE 17 MONTHS. CALCULATED TO RETURN IN 3,065 YEARS.
- GREAT COMET OF 1861. TAIL 24,000,000 MILES LONG, THROUGH PART OF WHICH THE EARTH PASSED ON JUNE 30TH. CALCULATED TO RETURN IN 400 YEARS.
- DONATI'S COMET OF 1858. TAIL 45,000,000 MILES LONG. CALCULATED TO RETURN IN 2040 YEARS.
- COGGIA'S COMET OF 1874. CALCULATED NOT TO RETURN TILL BETWEEN 6,000 AND 10,000 YEARS.
- GREAT COMET OF 1882. TAIL 60,000,000 MILES LONG. PASSED WITHIN 300,000 MILES OF THE SUN. VISIBLE 9 MONTHS.
- GREAT COMET OF 1843. TAIL 200,000,000 MILES LONG. APPROACHED TO WITHIN 538,000 MILES OF THE SUN. MAY RETURN IN ABOUT 400 YEARS.
- GREAT COMET OF 1910. TAIL 9,000,000 MILES LONG.

SPECIAL BASEBALL EDITION

SECOND EDITION

Financial

BOSTON AMERICAN

Weather Forecast — Continued Fair.

VOL. 7—No. 58. Registered in U. S. Patent Office. BOSTON, TUESDAY, MAY 17, 1910. Copyright, 1910, by the New England Newspaper Publishing Company. PRICE ONE CENT

ED IN ET

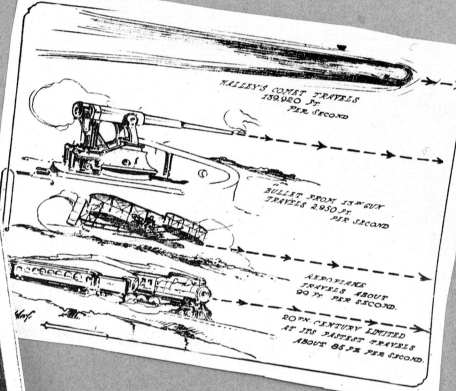

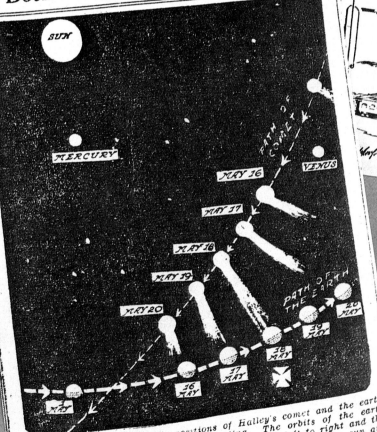

How the Comet and Earth Pass, Both Going 43 Miles a Second

The diagram shows the relative positions of Halley's comet and the earth during their approach, passing and parting. The orbits of the earth and the comet are shown, the earth moving from left to right and the comet slantingly from top to bottom, being now between the sun and the earth. They pass each other this evening, the passage beginning at 11:20 o'clock and ending about 1:20 o'clock to-morrow morning. At their relative speeds of eighteen miles a second for the earth, and twenty-five miles for the comet, they will pass at a speed of twenty-three miles an hour.

The old horse and buggy can't hold a candle to the Comet.

Comets come and Comets go,
They're such a sight to see.
They spread their tails
across the sky,
For all Eternity.

The Comet has captured all the holidays.

EXPERTS IN DOUBT AS TO EFFECT OF COMET ON EARTH

The constant change of form of Halley's comet, and its power to drop or grow a tail at will, as noted through their powerful modern instruments, have left various astronomers in doubt as to just what sort of a pyrotechnic display the heavenly visitor will make when Boston and the rest of New England celebrate "comet nights" to-morrow and Thursday.

NO HARM CAN COME TO THE EARTH.

While the students of the sky differ on many points regarding the comet, they are all practically agreed that no harm can come to the earth, even if we are brushed gently by the tail of the celestial flaming bird.

The earth is due to enter the 'tail' of the comet at 11:20 to-morrow night. Regardless of the fact that it is not believed that the comet will then be visible to persons in the eastern part of the United States, preparations are being made to look for it.

The Harvard Observatory has promised to inform fire headquarters in Boston when the comet is first seen in the western sky, and then the bells will ring a joyful notification so that every person in the Greater city may have a chance to see it.

It is expected that splendid views of the comet will be possible on Thursday and Friday evenings shortly after sundown, and it is expected that Boston and the rest of America will turn out in mass for that length of time at least, and the most interesting speculation was indulged in,

rushing head-on toward the earth at the rate of twenty-six and one-half miles a second, in an effort to make up the 13,000,000 miles between the two.

The comet is travelling at the approximate rate of 139,920 feet a second. The average projectile fired from a thirteen-inch gun travels 2,950 feet per second, a speed equal to about one one-hundred and twenty-fifth that speed of the comet. The fastest automobile in existence travels over a special course about 176 feet a second.

The Twentieth Century Limited, at its fastest, covers eighty-five feet per second. A gyroscope at full speed, under exceptional conditions, does 130 feet a second, while the fastest aeroplane ever built might make ninety-nine feet a second.

A trained sprinter, with a running start and the wind back of him, might do thirty-two feet in a second.

The earth, a pretty quick mover herself, is travelling toward the comet at the rate of eighteen miles a second. It has been calculated that if the earth and the head of the comet should clash they would come together with a speed several hundred times that with which a ball issues from the mouth of a cannon, and an energy 25,000 times as great, pound for pound.

READY FOR THE END OF THE WORLD IN THE BLUE GRASS

LOUISVILLE, Ky., May 18.—Preparations for the end of the world are being made to-day by the ignorant persons through the central and eastern parts of Kentucky. Whites as well as negroes think Halley's comet is bringing doomsday. Many farmers have refused to plant. All night revival services were held last night at Lexington and other places.

NEGROES FLOCKING INTO MEMPHIS, FEAR END OF THE WORLD

MEMPHIS, Tenn., May 18.—Thousands of negroes have flocked into Memphis the past twenty-four hours. They claim to have been told by a "prophet" that Halley's comet will destroy the earth at noon to-day. Special trains bearing the refugees have arrived from as far south as Vicksburg and as far north as Cairo, Ill.

Comet Pills And Other Things Hard to Swallow in the News

Anti-Comet Pills, $1 Per.

NEW YORK, May 17.—Third Officer John Retting of the Hamburg American liner Alleghany, in from West Indian ports, went ashore at Port-au-Prince, Hayti, and found a black medicine man peddling "anti-comet pills," guaranteed to stave off all malevolent effects of Mr. Halley's visitor, at $1 per pill.

Cheer the Comet
BY CLEMENT L. POLLOCK

PESSIMISTIC souls decry
    Halley's comet passing by
    And prepare, perhaps, to die
        In its tail.
But cheer up, dear girl and boy,
Banish phantoms that annoy;
It will leave a flood of joy,
        Not a wail.

Think of Summer's golden store,
In the mountain, on the shore;
Heaven lies right at your door;
        Don't repine.
For vacation time is near,
Field and forest green appear;
Say good-bye to grouch and tear;
        Fall in line!

Halley's comet wears a smile
Of good will, not to beguile
Dresses in celestial style
        With train to burn.
Then, Mr. Halley, here's a cheer,
We know we now have naught to fear,
Besides you will not find us here
        When you return!

COMET BLAMED FOR BIG FLOOD IN ITALY

ROME, May 16.—The river Tiber is swollen and floods to-day threaten many villages and towns. Snow is falling in the valleys of Lombardy and Tuscany, where the weather resembled the height of a bad Winter. Crops throughout Italy have suffered a great deal. The unusual weather has spread panic among the peasants who attribute it to the approach of Halley's comet.

HOST OF CONVERTS AT REVIVALS, ALL AFRAID OF COMET

PHILADELPHIA, May 18.—Halley's comet has brought a host of converts to revivalists. The Rev. Abraham Lincoln Johnson, a negro, is holding meetings on Monument road, West Manuyunk. His congregation last night was reduced to a paroxysm of fear as he pictured the mission of the comet and urged all to pray that the comet be warded from the earth.

Save us from the Comet!

The Comet's coming.

ABDUL FEARS COMET, REFUSES TO EAT

PARIS, May 20.—A despatch from Salonica to the Matin says that Abdul Hamid, deposed Sultan of Turkey, is in such fear of the comet that for several days he has refused to take food.

This Comet Person
By MORTON BIRGE.

The head of Halley's comet is growing larger. (Daily report.)

BEHOLD: the last thing the astronomers say
About the big comet that's coming our way—
It seems that puffed up with importance to-day
   Is the comet.

The scientists find that the comet has read
So much in the papers of what has been said
That all this publicity's gone to the head
   Of the comet.

Cheer up! Now we know that there isn't a chance
Of danger to us in the comet's advance.
It's doubtful if we get so much as a glance
   From the comet.

As you please, go ahead. We can stand it, you bet.
We'll struggle along a few centuries yet.
In fact, we are fussy who moves in our set,
   Mister Comet.

VENUS UNHARMED BY PROXIMITY OF SKY WANDERER

PROVIDENCE, May 16.—Frank E. Seagrave, an astronomer of this city, said today that the comet would be as close to the earth on Wednesday night as it was to Venus two weeks ago, and that he has taken a good look at Venus, and she has not been injured in the least.

"There seems to be absolutely no reason for fear of the comet," said Mr. Seagrave last night. "The comet is twice as far away now as it was in 1066 and no harm was done to the earth then. It will be just about as close to us Wednesday night as it was to Venus two weeks ago. I turned my telescope on Venus last night and she was still there. I couldn't notice that she had suffered any harm.

"It is a little hard to tell just how the sky will look Wednesday night. The comet's head will cross the sun's disk at 1:49 p. m. The sun will then be over the Pacific. The tail will be streaming on in the direction of the earth. I don't believe that we will have even a few shooting stars to testify to the proximity of the two bodies. If the earth were a little closer to the comet's head this night might differ."

The Comet has come!

HARVARD DEVICE SHOWS EARTH IS SAFE FROM COMET

Professor Robert W. Willson of the Harvard astronomical laboratory has constructed a small section of the solar system in a most ingenious way, by which he can demonstrate that there is no danger to the earth from Halley's comet.

"To talk of impending disasters," said Professor Willson, "is to talk of nonsense."

The device is a plate of glass raised on four pillows and pierced by a piece of wire that is bent at an angle of some 30 degrees. Through the centre of the plate is thrust a brass knob, which represents the sun and round it is traced a circle which represents the orbit of the earth.

Through this glass and reaching comparatively far above it at the extreme point of the arc is a piece of wire which indicates the course of the comet. It is a segment of the ellipse of some 3,000,000,000 miles over which the comet travels in its circuit of seventy-five years. The apparatus is accurately reduced to scale and the distance of 90,000,000 miles between the earth and the sun is the standard.

It shows that at the present appearance of the comet the only time when the comet will be directly in line between the earth and the sun will be on May 18. At all other times from now until some seventy-five years from now the comet will be either above or below the plane of the earth and the sun.

WE MAY OBSERVE AURORA BOREALIS, SAYS PROF. MILLER

PHILADELPHIA, May 18.—"The auroro borealis may mark the passage of the earth through the comet's tail, and wireless telegraph instruments may record slight electrical changes," said Prof. John A. Miller, director of the Stroul Observatory, Swarthmore, to-day. "This is contingent, however, on the tail's particles being charged with electricity. There may also be meteoric showers."

No 1910. Welt-Untergangs-Versicherungs-Karte. Inhaber dieses ist gegen den Untergang der Welt am 18. Mai 1910 mit **Zehntausend** versichert. Der Betrag ist nur spätestens 8 Tage nach dem Untergang der Welt bei der Halley-Bank zu erheben.

Die Direktion.

Komet 1910.

Trink' ma no a Tröpferl,
Trink' ma no a Tröpferl,
Eh' uns der da hat beim Schöpferl.

"Old woman, put away that umbrella. If the Comet sees your face, the Earth will be saved."

"I guess I'll wait out the collision up here".

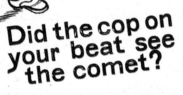

It will be Safer on the Moon.

*Ooh la la!
Before, during
and after.*

To the Moon by aeroplane, "inner tube", and balloon.

The Comet outwits the astronomers,
A dear one lost! Eternal regrets.

The Comet brings happiness.

Où est la Comète? — Wo ist der Komet?

HEY! LOOK OUT! THE COMET'S TAIL IS COMING FAST

THE COMET SCHEDULE.

Comet rose to-day.........3:08 A. M. Comet rises to-morrow....3:15 A. M.

Lats call for the comet! Mr. Halley's discovery got out of bed this morning at 3:08 o'clock, one hour and thirty minutes before the sun. To-morrow it rises at 3:15. After to-day the comet draws in toward the sun quickly, and on Wednesday evening passes between us and the sun. The earth will be enveloped in the comet's tail. After that date it will be in the evening sky, setting after the sun.

To-morrow the comet will be at the brightest, thirty-five times more brilliant than it was a month ago, and twice as brilliant as to-day. It will be seen in the west just after sundown.

For the next few days thereafter the celestial visitor, whom many Bostonians have become intimate with in the last week, should present a magnificent spectacle in the evening sky, with a tail pointing up from the horizon one-third of the distance to the zenith. Each night the comet will set later, getting fainter and fainter, until it bids us au revoir for another seventy-five years. This will be probably in the latter part of June.

Those persons who have been making their wills and bidding their families a tearful good-by in the expectation that the earth will plunge through the comet's tail to-day after to-morrow and be submerged

in poisonous gases will be relieved by the declaration of Professor David Edgar that the tail will approach no nearer to the earth than 13,000,000 miles.

"The comet's tail and the earth are now 20,000,000 miles apart," said Professor Rice. "Although they are moving head-on at a relative speed of forty miles a second, I am sure they will not touch. No doubt the astronomers have predicted that they will meet. In fact, many astronomers believe the tail which we see is only an optical illusion caused by the sun's reflection."

"The earth on May 18 will be 93,000,000 miles from the sun. The comet will be between the earth and the sun—13,000,000 miles from the earth."

Despite Professor Rice's soothing statement, astronomers all over the country are still figuring that the tail will sweep over the earth on Wednesday.

Halleys' comet will make its last morning appearance to-morrow morning, when it will peer over the horizon at 3:14 o'clock. It will be invisible until daylight.

Continued on Page 2, Column 5.

FIRE ALARMS TELL OF COMET'S COMING

The fire alarm will be sounded in Boston to-morrow morning for Halley's comet. Fire headquarters will be in readiness about 3 o'clock to get word from Harvard observatory of the appearance of the comet. The word from the observatory will be the signal for fire bells connected with the city alarm system to sound. Mayor Fitzgerald this afternoon made arrangements, at the same time notifying the public.

No person in Boston, New England or the whole world, for that matter, need fear the comet, whose fuzzy tail will brush the earth gently this week. The big astronomers of Harvard University, and of other places agree in that in statements made to-day.

While the comet will pass between the earth and sun about 10 o'clock on Wednesday night, and conditions are ideal for the trip, Professor E. C. Wendall of the Harvard observatory said to-day that Boston will not catch a glimpse of it because the sun has long since gone down in this section of the country. The people on the Pacific slope may see it and in Japan and the Philippines they will have a fine view of it.

Bostonians should get their best view of the comet on Thursday or Friday evening in the western sky. On those evenings the Halley visitor should make a brilliant display.

May Be Shower of Dust.

"No one need fear that the comet will harm us," said Professor Wendell to-day. "There may be a shower of brilliant dust in the sky, but it will not hurt anything. People need not worry.

"In the first place the comet will be nearly 15,000,000 miles away at its nearest point, and that is too far away to do any damage. The comet is now about 20,000,000 miles away, so it is traveling pretty swiftly toward us at this time.

"We were able to get a glimpse of the comet at 3:15 a. m. to-day, but it lasted just about a minute. We do not expect to see it again until Thursday or Friday evening. Then it will be in the west and should be perfectly visible to the naked eye. It will gradually fade after that."

Professor Percival Lowell, who returned from Europe, reaching Boston to-day, declared that the tail of the comet is so vacuous that it will not hurt anything. His description of the flaming visitor is that "it is the airiest approach to nothing set in the midst of naught."

Professor Robert W. Willson of Harvard University, has made a contrivance showing the relative position of various heavenly bodies including the earth, comet, sun and so on, by which he quickly demonstrates that there can be nothing feared from the comet. He says that there may be a shower of charged particles that would give pyrotechnical display without damage.

Some of the astronomers suggest that the comet may affect wireless telegraphy, but that they do not believe it will influence anything else on the earth.

Those persons who have been making their wills and bidding their families tearful good by in the expectation that the earth will plunge through the comet's tail the day after to-morrow and be submerged in poisonous gases will be relieved by the declaration of Professor David Edgar that the tail will approach no nearer to the earth than 13,000,000 miles.

20,000,000 Miles Apart.

"The comet's tail and the earth are now 20,000,000 miles apart," said Professor Rice. "Although they are moving 'head-on' at a relative speed of forty miles a second, I am sure they will not touch. No bona fide astronomers have predicted that they will meet. In fact, many astronomers believe the tail which we see is only an optical illusion caused by the sun's reflection."

"The earth on May 18 will be 93,000,000 miles from the sun. The comet will be between the earth and the sun—13,000,000 miles from the earth."

Despite Professor Rice's soothing statement, astronomers all over the country are still figuring that the tail will sweep over the earth on Wednesday.

Halleys' comet will make its last morning appearance to-morrow morning, when it will peer over the horizon at 3:16 o'clock. It will be invisible until daylight.

Continued on Page 2. Column 2.

WHOLE SCIENCE WORLD WAITS COMET'S TAIL AS IT SWEEPS EARTH

Continued From First Page.

but only faintly, as the moon will be bright and its light will dim the rays of the comet. On Wednesday morning the comet will still be in the East, but it will not rise until 4 o'clock, when daylight will have come. Again on Wednesday, when it will have chasseezed across the space between the earth and the sun, it will be visible in the evening skies, but still under the diluting influence of moonlight, as the moon fulls on Wednesday.

On Friday and Saturday of this week the latter rising of the moon will furnish a better opportunity for evening comet seekers. It will be visible on Friday at 8:46 p. m. and will remain in sight until 10:14. On Saturday evening it will be visible from 9:33 until 11:47.

The evening reckonings are all predicted upon the supposition that nothing will happen on Wednesday, when the comet's tail sweeps the earth, to deprive dwellers of this sphere of the desire to see the comet. Scientists are still figuring out what may happen. Most of them are sanguine, but some of the astronomers of France, who are held in high esteem by those of other nations, refuse to positively say that the sweeping of the tail will not be fatal to all animal life on the earth.

Pickering Has no Theory.

Professor E. C. Pickering, director of the Harvard observatory, says that he has no theory as to the constitution of the comet's tail, but he believes that no effect of its passing will be perceptible on the earth. The professor will be watching for any unusual phenomenon when the tail sweeps by.

Professor Robert W. Wilson of Harvard, and Professor David W. Todd of Amherst, agree in the belief that there may be a darkening of the earth's atmosphere, followed by meteoric showers or an auroral illumination. Neither has any fear of cyanogen gas poisoning the atmosphere. Professor Wilson thinks that the wireless telegraph service may be affected a trifle.

Professor S. A. Mitchell of Columbia University, says that it is likely that nothing unusual will happen when the earth passes through the comet's tail. The rarity of matter in the tail is such in comparison with the solidity of the earth's atmosphere, he says, that the tail will not take away any of the atmosphere with it. No harmful effects need be feared, he says.

W. W. Campbell, director of Lick observatory, says that nothing need be feared on Wednesday. There may be a few phenomena, such as meteorite showers or auroral glows, but these will be of no great importance.

No Data to Go By.

J. M. Schaeberle, professor of astronomy at Ann Arbor University, Michigan, says that scientists are without sufficient data to make a definite prediction of the phenomena that will occur in consequence of the earth's passage through the tail.

Professor H. N. Russell of Princeton says that the apparent motion of the tail through the sky will be the only observable effect of its passing. It is of fine gaseous or solid particles, he says.

J. A. Brasliear, director of the Allegheny Observatory, doubts any physical effect on the earth. The tail is so attenuated, he says, that he has often viewed stars through 200,000 miles of the tail without finding their brilliancy dimmed.

David Gill, president of the Royal Astronomical Society of London, doubts if the tail is long enough to reach the earth, and is sure that even if it is it will do no harm.

Michael Giacobin, astronomer at the Observatory of Paris, declares that the passage of the tail ought to be a source of felicitation that a magnificent spectacle is to be seen, rather than a source of fear.

Streamers in the sky are predicted by Andrew C. D. Crommelin, the noted English astronomer.

May Affect Atmosphere.

Henry Des Landres, director of the observatory at Meudon, France, says that there is no absurdity in the hypothesis that the presence of the comet may have a perturbing effect on the earth's atmosphere. "This we can neither affirm nor deny," he says.

The naval observatory at Washington is making the most complete preparations for the observation of the comet's passage. The professors at the observatory disagree as to the probability of meteoric showers, but all agree that there will be nothing harmful to mark the phenomenon.

Sir Robert Ball, the eminent British astronomer, gives the warning that Wednesday's opportunity will not be repeated for seventy-five years, and that every one should try to see the comet. He points out that a pair of binoculars will give a better view of it than a telescope.

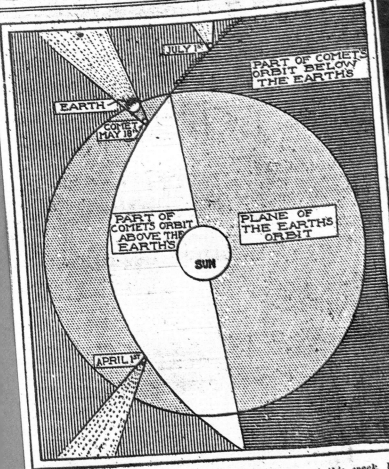

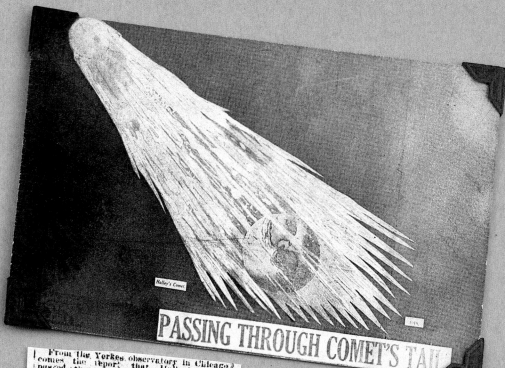

PASSING THROUGH COMET'S TAIL

GREAT COMET IS LOST

Halley's comet is really and truly lost. After two days of heart-breaking search by a posse of almost a million men armed with telescopes, the celestial wanderer which was to light up the world and lots of other good things has vanished.

Professor O. C. Wendell and Professor E. C. Pickering of the Harvard observatory were almost heartbroken to-day. After four months of ceaseless labor in preparing for the welcome to the comet when, this week, it should make its call on the earth, they have been disappointed. They did not see even a speck of its two tails.

The astronomic feast was spoiled. It is like a funeral without a corpse.

The far West, which expected to gloat over Harvard and other eastern astronomical observatories, has had no better luck. Reports of the tail come from the West, and so are that whether it is asking passed through the tail of the comet. Said Professor Wendell of the Harvard Observatory to-day:

Think Tail Lagging Behind.

"We don't know what happened to Halley's comet. It may have passed the earth last night without the earth entering the tail in any way. If it did pass, it did not come very close to us.

"The only solution we can get for it is that the tail is behind time—lagging far behind the nucleus. This theory is borne out in part by the report from the observatory at Berkeley, Cal., that Professor Leushner there saw the comet on the 18th and found the tail lagging fifteen degrees behind the head; that would be a distance thirty times the diameter of the moon. It may not have caught up yet.

"Some persons reported to me to-day that they saw traces of the comet at 3 o'clock this morning. Several of the astronomers at Harvard also saw a band of light passing the square of Pegasus. I do not believe that this was in any way due to the comet. The location is not right. While we have lost track of the comet for the past two days we feel that it is still following its course and would not be in the location described.

"The Rev. J. B. Metcalf sent word to the observatory that he saw the comet at 3. We have no details to connect his observation with those we made."

May Be Seen To-night.

"If the comet appears to-night it will show up about one hour and twenty-eight minutes after sunset. That brings it somewhat out of the strength of the sun's rays. It should be observable if the clouds do not interfere.

"Later, early next week, it will be visible if conditions are favorable later in the evening and with less moonlight to bother the vision."

From the Yerkes observatory in Chicago comes the report that Halley's comet passed the sun's disc on schedule time, about 9 o'clock Wednesday night. It contained comparatively little solid matter. Its tail, looped like a twenty million mile question mark, did not pass across the earth.

A concensus of scientific opinion to-day fails to furnish an explanation for this amazing and unlooked for phenomenon, but was thoroughly agreed taht the comet's tail had suffered a gigantic curvature.

Astronomers Are Divided.

"Did we pass through the tail yet, or are we passing through it now?" These are questions upon which astronomers are divided in opinion.

"There will be absolutely no way to tell when the transit takes place until we see the comet in the west. I believe, though, that we will see it as the sun sets Friday," said Director Frost of the Yerkes observatory.

"It is quite possible that we may be in the tail of the comet now, and equally possible that we may not pass through it at all because of the sudden curve it has developed," said Prof. Frost. "If we are in the tail there is no way of telling it because the tail cannot be seen and probably produces no effect on the earth at all. These clouds are very disappointing, as we had hoped to solve the riddle this morning."

Believes It Very Small.

"Dr. Hale's observation would seem to indicate that we have not yet reached the tail and that its curve away from the earth is still more pronounced.

"This is nothing uncommon with comets, however, because their tails often turn and twist or even break up with surprising rapidity.

"The observations of the transit confirmed the opinion of the best authorities that a comet has really a very small mass indeed. The nucleus of Halley's was not large enough and solid enough even to be visible as a black dot crossing the sun. This confirms my belief that the earth could collide with a comet of the size of Halley's and suffer no ill effects. The whole comet, if compressed, could be carried away in a freight train and a comet of that size striking the earth probably would result in a brilliant shower of meteors and nothing more."

NO COMET PERIL FOR THE EARTH, SAY EXPERTS ABROAD

LONDON, May 16.—There is no danger to the earth or the life upon it, animal or vegetable, in Halley's comet. This is the concensus of opinion among European scientists.

The majority of them declare that it is highly improbable that there will be any perceptible effect when the earth speeds through the comet's tail for two hours on Wednesday night.

Extraordinary preparation for the observation of Halley's comet, the most famous of the erratic members of the solar system are being taken throughout Europe. The advance of science in the last seventy-five years, or since the comet last whirled by the earth, is illustrated in the present-day facilities for its study.

At Paris, London and Berlin astronomers, in the next four days, will make ascensions by balloon to study the visitor from the upper air. Thousands of telescopes will be trained upon it and the spectroscope and all the other amazing instruments of the twentieth century astronomer will be brought into play at close range. If the tail of the comet extends 15,000,000 miles beyond the head it will sweep over the length of the earth. Estimates will place the tail propelled away from the sun by the force of the latter's heat at from 20,000,000 to 46,000,000 miles.

The passage through the tail begins at 11:30 p. m. Wednesday. The comet will be nearest the earth, distant then only 14,300,000 miles.

The length of the tail has fluctuated to scientists' estimates. According to Senor Miquez of Madrid it lengthened from 23,412,000 miles to 49,981,000 in a few hours on May 13 and 14. Senor Miquez agrees with many others that the tail contains no deadly cyanogen gas.

HELP! EARTH HAS DESIGN ON COMET'S TAIL

WASHINGTON, May 18.—The earth may steal part of the comet's tail. Willis L. Moore, chief of the Weather Bureau, today said that special orders had been sent to weather men throughout the country to watch for such a development which would be apparent in the color of the sky, particularly around the sun and the moon.

"That part of the tail may be retained is not at all unlikely," he said. "We do not anticipate anything extraordinary tonight, while passing through the tail, however. There may be electrical disturbances, though, and we may feel electro-magnetic effects. If the comet's tail is composed of particles of dust, the effect will be optical, for the most part."

Professor Asaph Hall, chief astronomer of the United States Naval Observatory, said:

"The majority of scientists do not believe that the passage of the earth through the tail of the comet to-night will be accompanied by any particularly noticeable outward manifestations. In the case of meteoric display or any aurora we will take observations and photographs. In China, the actual crossing will take place in broad daylight. For that reason oriental scientists will have much the better opportunity for observing the movement."

Professor Hall added that the reason why the passage will best be seen in China in daylight is because of the fact that when the comet passes between the sun and the earth it will be possible to observe it with the telescope.

"Passing across the disk of the sun," he said, "it stands to reason that it will be observable at that time if the weather conditions permit, while in countries where it is night, there will be no sun seen. The passage will be equally observable in Japan, the Philippines Islands and all that part of the world. In this part of the globe, of course, the actual passage will not be seen to-night. The only evidence that may be observed may be an aurora, but it is impossible to tell if even this will be apparent. In short, scientists, like laymen, are absolutely unable to say what will or will not be seen."

The Weather Bureau has instructed its observers throughout the United States to make careful record of anything exceptional. They are instructed to look out for meteors.

THE COMET.
By Oliver Wendell Holmes.

THE comet! He is on his way
   And singing as he flies;
The whizzing planets shrink before
   The spectre of the skies;
Ah! well may regal orbs burn blue,
   And satellites turn pale,
Ten million cubic miles of head,
   Ten billion leagues of tail!

On, on by whistling spheres of light
   He flashes and he flames;
He turns not to the left nor right,
   He asks them not their names;
One spurn from his domestic heel—
   Away, away, they fly,
Where darkness might be bottle up
   And sold for "Tyrian dye."

And what would happen to the land,
   And how would look the sea,
If in the bearded devil's path
   Our earth should chance to be?
Full hot and high the sea would boil,
   Full red the forests gleam;
Methought I saw and heard it all
   In a dyspeptic dream;

I saw a tutor take his tube
   The comet's course to spy;
I heard a scream—the gathered rays
   Had stewed the tutor's eye;
I saw a fort—the soldiers all
   Were armed with goggles green;
Pop cracked the guns! whiz flew the balls;
   Bang went the magazine.

I saw the scalding pitch roll down
   The crackling, sweating pines,
And streams of smoke, like waterspouts,
   Burst through the rumbling mines;
I asked the firemen why they made
   Such noise about the town;
They answered not—but all the while
   The brakes went up and down.

I saw a roasting pullet sit
   Upon a baking egg;
I saw a cripple scorch his hand
   Extinguishing his leg;
I saw nine geese upon the wing
   Towards the frozen pole,
And every mother's gosling fell
   Crisped to a crackling coal.

I saw the ox that browsed the grass
   Writhe in the blistering rays,
The herbage in his shrinking jaws
   Was all a fiery blaze;
I saw huge fishes, boiled to rags,
   Bob up through bubbling brine;
And thoughts of supper crossed my soul;
   I had been rash at mine.

Strange sights! strange sounds! O fearful dream!
   Its memory haunts me still,
The steaming sea, the crimson glare,
   That wreathed each wooded hill;
Stranger! if through thy reeling brain
   Such midnight visions sweep,
Spare, spare, O spare thine evening meal,
   And sweet shall be thy sleep.

HALLEY'S COMET, JUNE 1910.

La Comète 1910

Behind my house, alone at night,
A Comet in the sky I sight.
I wonder what it really is
— a ball of ice and tail of fizz?

HALLEY'S COMET

Photographed at the Transvaal Observatory,

Komet 1910

BAD WEATHER HIDES COMET IN THE WEST

Professor Pickering of Harvard observatory to-day received the following letter from Professor Edwin B. Frost, director of the Yerkes observatory in Wisconsin:

"If as unfavorable weather has prevailed in the rest of the country as has been the case here during the last ten days, the following observation of Halley's comet by Professor Barnard on Sunday morning, May 15, will be of interest:

"A heavy bank of smoke and haze obscured the eastern horizon, but Professor Barnard was able to trace the tail with the naked eye to a distance of fifty degrees from where the head would have been seen. The tail was three and one-half degrees broad at three-quarters of its length from the head. Later, in his five-inch guiding telescope, he found the nucleus still stellar, but of a very small fraction of the brightness of Venus seen under similar conditions of haze.

"We have been unable to secure any satisfactory spectrogram since the morning of May 7, on account of the distressing conditions of the weather."

To Escape the Comet, Hire Submarine Boat

Deadly cyanogen gas does not travel through the water. So to escape the comet hire a submarine boat, fill it with three days' edibles and drinkables, and go under water to-morrow. The deeper the spot the better.

Stay under for three days, not even poking your nose above for one moment. At the end of that time, if nothing happens to your submarine, the cyanogen gas in the world will have spent itself.

Then, if all the people in the world have perished under the deadly gas, you can claim the world for your own.

If they are not dead—perhaps you can stand the laugh.

Halley's Comet 1910 May 6
Photographed at the Southern Station of the Lick Observatory

JUST HEAVY RAIN; THAT'S ALL COMET WILL DO TO EARTH

PARIS, May 18.—The most eminent French astronomers declared to-day that Halley's comet has no cyanogen gas, but heavy torrential rains are expected, which will be followed by a long period of drought.

BEST COMET PHOTOGRAPH

This photograph was taken last week at the great Mt. Lowe Observatory in Southern California. It shows Halley's comet as the great wanderer is approaching the earth at tremendous speed.

TAUNTON MINISTER HAS GOOD VIEW OF HEAVENLY BODY

TAUNTON, May 18.—Observations made at 2:40 a. m. to-day by the Rev. Joel H. Metcalf showed the length of the comet's tail to be 120 degrees and the width 11 degrees. The tail extends from the star Gamma Pagasi to Theta Aquilla. It is nearer the earth's ecliptic than it was yesterday, showing that the sweep of the tail in the movement of its transit was well under way. This movement continuing will cause the tail to sweep the earth to-night. Conditions were favorable, there being only a few light clouds here.

By TAD.

FIFTH AVE. AND BROADWAY GARY, IND. MAY 20TH 1910.
SHOWING HALLEY'S COMET.
COPYRIGHT 1910 BY

DE KOMEET VAN HALLEY.

Als de staart van deze komeet langer is dan 23 000 000 K.M. zal deze in den nacht van 18 op 19 Mei 1910 den dampkring der aarde raken. Men behoeft echter niet beangst te zijn, daar er ten hoogste een sterrenregen het gevolg van kan zijn.

Admire the Comet; Don't Fear It

By ELLA WHEELER WILCOX

Copyright, 1910, by American-Journal-Examiner

THE great comet will now be visible for several weeks in the evening sky.

Thousands of nervous individuals have been encouraging superstitious fears of this brilliant visitor; many people who are sensible and sane in mind, on most subjects, have felt apprehensive of disaster regarding the comet; and almost every one who possesses temperament has been stirred by the thought of it. It is quite natural that such an event in the solar system should affect human beings.

The elements are all affected by the movements of the planets. Certain conjugations of stars produce certain kinds of weather; and when the planets speak, the tides and the oceans respond.

Man is a part of the Universe; he is composed of exactly the same chemicals which form the stars, the earth, the oceans and space. What more natural than that he should feel any great event which occurs in nature.

Just as all nations feel, to some extent, any great crisis in the political, or financial, or social world, of another nation, so man feels what is happening in the solar system.

The comet is a periodical disturber of solar order and regularity.

It is like a fiery, oratorical revolutionist, or an evangelist preacher who resorts to spectacular methods to obtain the attention of mankind, and to stir people out of their settled ways and ideas, and, like the revolutionist and the evangelist, it always succeeds in waking fear; and sometimes it produces such excitement and agitation that war is the result.

It is no evidence of weak superstition to think the comets are heralds of war.

Comets do not come to announce war. But the coming of the comets produces electrical effects upon human organizations which frequently result in war.

This result can be avoided by the cultivation of self-control of individuals; by sensible reasoning, by philosophical analysis and an understanding of cause and effect.

A number of excitable people committed various crimes and several suicides occurred, attributable to fear of the comet.

Now as the fiery visitor is departing, rid your mind of all fear. Think of the comet as a GREAT BALL OF RADIUM which is sending out rays of magnetism to give you health, happiness and power. Believe that it is a battery of electric force which will recharge all your physical, mental and spiritual machinery—if you make yourself quietly receptive.

Remember you are composed of EXACTLY THE SAME MATERIALS which make the comet; and that you are therefore its kin, and that the same STUPENDOUS CAUSE put you upon your orbit, which flung that comet into space to make its periodical rounds.

It can give you strength, light, inspiration and hope, if you open the door of your being to such rays.

As you look at it, say the names of such qualities as you desire. Cast out all fear. Admire and reverence the spectacle.

It is a part of the magnificent universe of which you, too, are a part. The comet has come to rouse the indolent into action. Let it help you to be strong with love, health, wealth and power for usefulness.

EARTH READY TO ENTER TAIL OF THE COMET

To-morrow night, at 11:20 o'clock, Boston time, the head and tail of Halley's comet, the sun and the earth will be on a direct line. The tail of the celestial visitor will stretch out toward the earth, and its 15,000,000 miles or more of length may whisk the earth. Will the tail be long enough to reach us? is the question every learned scientist is asking now.

"It will," says Professor Mitchell of Columbia.

"It won't," says Professor Gill of the British Royal Astronomical Society.

"It may or it may not," says Waldemar Kaempffert of the Scientific American.

Wild Guesswork, Says Scientist.

"In fact," added Mr. Kaempffert, in an interview with an AMERICAN reporter, "all the men who are venturing opinions as to whether or not the tail of the comet will reach the earth are simply indulging in wild guesses. If the tail remains its present length, we shall pass through it about 8,000,000 from the end. But the tail may shorten millions of miles in a day. Its thin substance is variable.

"On February 14 last the tail was 15,000,000 miles long. A few days later there was no tail at all. Subsequently the tail was regained, and it is still trailing the comet—for how long no one knows. The comet's tail didn't touch the earth in 1835, its last trip, and it is in almost the same position now as then.

"As Professor Cromelin of the Greenich Observatory points out, the tail is composed of such tenuous gas that the suggestion that it might contains meteors seems quite impossible of entertainment. The fact is, astronomers are not in a position to predict with any degree of certainty just what phenomena will result from the passage of the earth through the tail. But it can be stated with positive assurance that the comet is not a menace to life on the earth."

Boston Will Sit Up.

Nevertheless, laymen want to be "shown." Although astronomers declare ther will be no visible effects if the earth passes through the tail, Boston is preparing to be out of dors at 11:20, the hour when the earth is due to enter the tail, to see what will happen. Points along Beacon Hill and other high places will be preempted at an early hour, and comet parties will be in order on the roofs of hotels.

Professor William R. Brooks, director of Smith's observatory at Geneva, N. Y., who took recent observations of the comet, reports that the tail is now stretching out in a broad band across the sky to a length of more than forty-five degrees. The nucleus, Professor Brooks says, was very bright, being easily visible with the naked eye long after all the stars except Venus had disappeared.

New York Comet Night.

"Comet Night" in New York is to be the importance given to New Year's eve. On the roofs of hotels and cafes there are to be innumerable "comet parties," and if disaster is to come it will find the big Metropolis in merry mood.

People in other parts of the world await the event with different emotions. Already a score or more of suicides due to fear of the comet have occurred, and many persons have become crazed with the excitement of anticipation. In the British Isles and in Europe the superstitious attribute the death of King Edward VII. directly to a malign influence attending the comet on its visitation.

THE WORLD'S WORK ADVERTISER

Pears' Soap

Halley's Comet appears but once in seventy-five years. Pears' Soap is visible day and night every day of the year all over the world and has been since 1789, in the homes of discerning people.

OF ALL SCENTED SOAPS PEARS' OTTO OF ROSE IS THE BEST

"All rights secured."

In writing to advertisers please mention THE WORLD'S WORK

IN TAIL OF COMET NOW, SAYS TODD

The earth is probably ploughing its way through thousands of miles of the tail of Halley's comet to-day, according to Professor David Todd of Amherst College. Instead of being one long projection, as was expected, the tail is divided into great parts, one of which is curved and sweeps for a few million miles away from the sun, toward the direction in which the earth is approaching the comet to-day.

The clearest view yet obtained of the comet was had by Professor Todd early to-day from his observatory. The tail was over seventy-five degrees long—a distance equal to fifty times the diameter of the moon. The head or nucleus was not visible, being dimmed by the brilliancy of the sun.

"When the earth passes through the comet's tail, and it may be doing it in part to-day," said the professor, "the experience will be no more startling than putting your finger through the beams of a searchlight.

May Only Graze Tail.

"Just how far we shall travel through the main tail of the comet to-morrow we cannot say definitely. It will probably be about 150,000 miles if we go straight through it. But we shall probably only graze it, and that will make the distance much less.

"Early to-day the brilliance and length of the comet's tail was surprising. With the head—if the head had been visible—the total length was about eighty degrees, which would be about 160 times the diameter of the moon. It has not been so large before.

"The tail was divided into two parts, one faint and the other quite brilliant. The one pointing north was the brighter. The other is the one we may be passing through to-day.

First Case of Double Tail.

"Between the two tails, for they were really two, was a dark space of sky probably three degrees wide. As I remember it this was the first distinct picture of a double tail that I have seen thus far.

"It will be worth while staying up to-night. The sun may eclipse the main part of the comet in the morning, and there may be no meteorites or shooting stars. But there is a chance that the tail of the comet will flare up and make a picture like the Northern lights.

"To-morrow night will be the great time for the comet. It will be better in the Eastern hemisphere than here, but there is a chance that some grand illumination will take place. We do not know exactly what to expect, but we are going to watch every minute and not miss anything.

"Next Monday, if the sky is clear, will be the best time to view the comet. It will be visible in the evening. The moon will be eclipsed and the comet will have the whole heavenly field to itself, with no sun or moon to outshine it."

WHERE WILL IT STRIKE NEXT
A Comet That Has Cut Loose in the Republican Constellation

The appearance of a Comet in the Sky has been taken as a powerful omen in the political affairs of mankind.

THE NEW COMET—A PHENOMENON NOW VISIBLE IN ALL PARTS OF THE UNITED STATES.

In the year 1870, politicians and labor leaders focused on what they regarded as a new threat. Thomas Nast saw it as a Comet in this cartoon he did for Harper's Weekly of August 6th.

Brainstorm Tips on the Comet by Direct Wireless

"Will the comet's tail injure the earth?"
In an effort to get some light on this burning question a reporter for the AMERICAN indulged in a "brainstorm" to-day, and when the mental clouds had passed away he announced that he had gotten in tune with the infinite, and received communications from noted persons on various planets, the moon and the sun. Likewise he produced some wireless pictures. Following is the result of his efforts:

By C. FARANDKEEN.
(The Wisest Man on the Planet Saturn.)

From this planet it would appear that the tail of Halley's comet, as it approaches the earth, is being outshone by Comet "T. R. Africanus." This comet was first noticed in the constellation U. S. A. It then shot into Africa; thence it veered into Europe

and soon it is to return to the spot whence it started. During its trip it was observed to shoot forth much flaming gaseous matter. While this comet has been blazing forth I have also observed a black spot in California. It is bearing down on a meteor, which has been named Jeffries. The question now agitating this planet is: "What will be the result of the collision?"

By PROFESSOR C. ANT. COMEIT.
(Who worked himself up from the position of canal-digger to that of Astronomer Extraordinary and Guesser A-Plenty on the Planet.)

I went out in my airship this morning and hailed the comet. It hailed back at me with redhot stones. Said I to the comet: "Will your tail injure the earth?"

"Alas!" it replied, "there isn't a chance. My tail, you know, is made of burning gas, and when that comes in contact with the watered stocks in State street it will be goodby tail." It then sent a jet of laughing gas at me and I was projected into a canal. I ought to get a new suit of clothes and a heavy expense bill for this.

By COLONEL FIRE-EATER.
(President of the Empire of the Sun.)

I want to say right here that I don't give a continental flash of light about

Halley's comet. What I want to know is what show the Tigers have got to win the pennant after being slugged by Pat Donovan's bunch? Also, can the Dov keep up their speed?

Our Own Comet

"There They Were Waiting at the Pier."

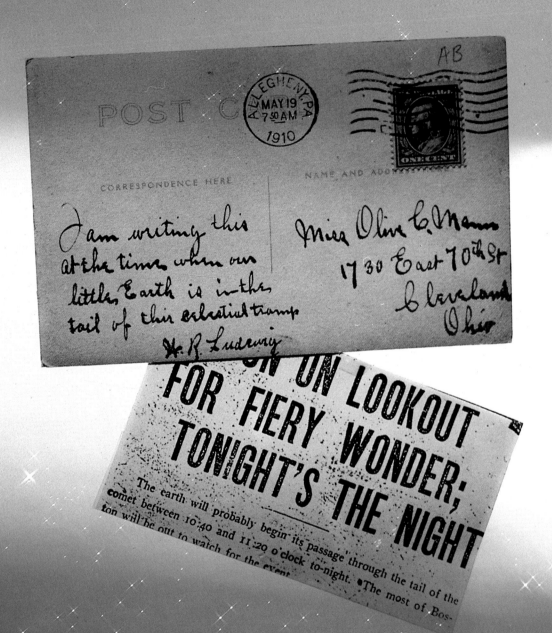

THE COMET AT LAST!

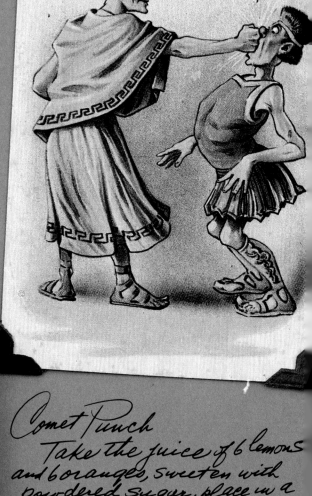

Roman Punch

Comet Punch
Take the juice of 6 lemons and 6 oranges, sweeten with powdered sugar, place in a large punch bowl with a block of ice, and add: 1 quart of brandy, 1 quart of sherry, 4 quarts of champagne, 2 quarts of sparkling water. Stir well and decorate with fruits of the season.

HOW BOSTON IS TO VIEW THE COMET

AMERICAN reporters to-day asked prominent Bostonians: "What are you going to do to-morrow, Comet Day?" The answers range all the way from the very serious to the very funny. They show that Bostonians are not afraid of Mr. Halley's comet. Here are the answers:

GOVERNOR DRAPER—"I have no more fear that the comet or its tail will do harm than I have that the Democratic party will be successful next Fall."

LIEUT.-GOVERNOR FROTHINGHAM—Having been through several political campaigns, I don't imagine that a little thing like the tail of Halley's comet will bother me to-morrow.

MAYOR FITZGERALD—The near approach of the comet to-morrow isn't worrying me much. In fact, I've got worries enough on this earth without wondering about outside influences.

COLONEL DICK FIELD, the Mayor's confidential man, said—After bucking the Finance and Civil Service Commissions, one doesn't fear a comet. This comet comes for a day and then stays away for about a century. We've got some worries about the Mayor's office which are with us always.

UNITED STATES MARSHAL GUY MURCHIE—I hadn't thought much about it. We have comets of our own around here this week.

FIRE COMMISSIONER PARKER—I am going to stick with the department. It may need me.

REV. DR. GEORGE W. KING of the People's Temple—Inasmuch as there are 17,000,000,000 other comets in the heavens I'm not worrying over this one. It will come again and see us, it may not be in my life time, and I do not believe it will do us any harm May 18.

An Undertaker's View.

LEWIS JONES—I do not know anything about comets, but I intend to be happy and contented the day that it gives the earth a smile. I am not looking for the comet to kill off people and make it a busy day for me and the astronomers.

WILLIAM HODGE, actor—I don't know what will happen; I guess a part of the tail hit my cast last night, as one of them has appendicitis.

EX-PENAL COMMISSIONER JOHN B. MARTIN—I have read with interest the testimony of scientists in regard to the coming of Halley's comet and its effect in purifying the atmosphere. Let us hope it will purify the political atmosphere of the Old Commonwealth and give to the people something more revolutionary than the comet to do this. The real estate market will give a cordial welcome to the comet.

COMET NIGHT TO DRAW GREAT THRONGS TO HIGH SPOTS ALL OVER BOSTON

Guild Gets Out of Way.

EX-GOVERNOR CURTIS GUILD—I don't believe the comet can strike near me to-night, for I will be speeding by train from Greenfield to Boston. I am to speak there to-night, urging the young men to become interested in the national guard.

MARTIN LOMASNEY—The members of the Hendricks Club have tried to read some political dope in the tail of the comet, but I don't think they have agreed upon anything yet. Some of the boys may stay up to see it to-night, but yours truly will be asleep.

HUGH JENNINGS, manager of the Detroit Baseball Club—The Tigers are studying inside baseball, not comets, just now. I think the boys will play just as good ball whether the comet comes or goes.

"Votes for Women" as Rival.

MISS MARGARET FOLEY, of the Boston Equal Suffrage Association—If there was a big 'V. W.' kite up there with it to rival the thing in interest, that would be something like. I would rather have a good kite than a comet any day. You can pull a string on a kite but a comet simply glides along and holds a monopoly on everything. I would like to make it 'ring off' on public interest.

J. J. MARTIN, banker—The comet looks to me like a cat's tail and I'm no more afraid of it than I am of a cat. The day when it gets near to us will find us at business, and it cannot affect the steadiness of the real estate market for it will take something more revolutionary than the comet to do this. The real estate market will give a cordial welcome to the comet.

COMET PARTIES TO MAKE BOSTON GAY TO-MORROW NIGHT

Have a cocktail on the comet.

Comet parties are being organized for to-morrow night in all the best restaurants and hotels in the city. Drinks are being coined for the occasion, tables reserved, and extra decorations put on. If you don't go to-morrow you won't get another chance for about seventy-six years.

"Billy" Smith of the Hotel Hayward has made arrangements for six comet parties. Half the tables are reserved, and the demand for more is expected to swamp the place before to-morrow.

The American House rathskeller is making reservations. The Boylston cafe plans a real comet night for those who come early enough to get seats. All the "Bohemian" places are planning especially for it. It is expected that the roof of the Westminster will be the centre of a dozen comet parties. It will be an ideal place to see what is to be seen at 11:20 in the evening when the earth passes through the comet's tail.

The comet cocktail is so far the most promising concoction for the evening. Its ingredients are not given out, but it is known that it will be longer and more effective than the ordinary drink.

Then there will be other drinks. Halley's highball won't last seventy-six years, but it is guaranteed to last for some time. The Nucleus brandy will offer a sunbeam in every sip. A cyanogen flip is guaranteed to do more damage than the comet's tail. Also perhaps to cause as much merriment.

The hotels will be crowded with the after-theatre crowds. The promise by the city to ring the fire bells on notice from the Harvard observatory will draw thousands down town to watch, and while watching to while away the time with some of the especially prepared programmes.

Remember, it's your last chance for seventy-six years!

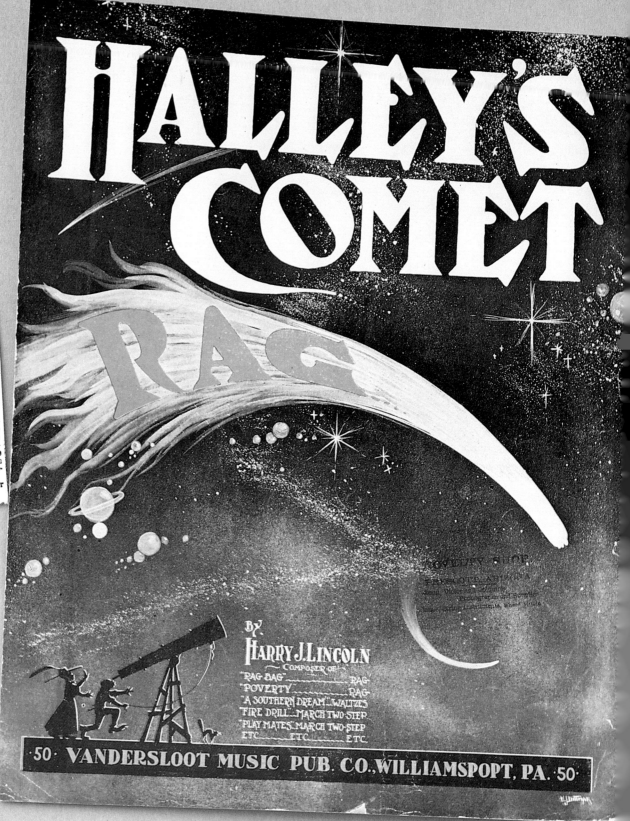

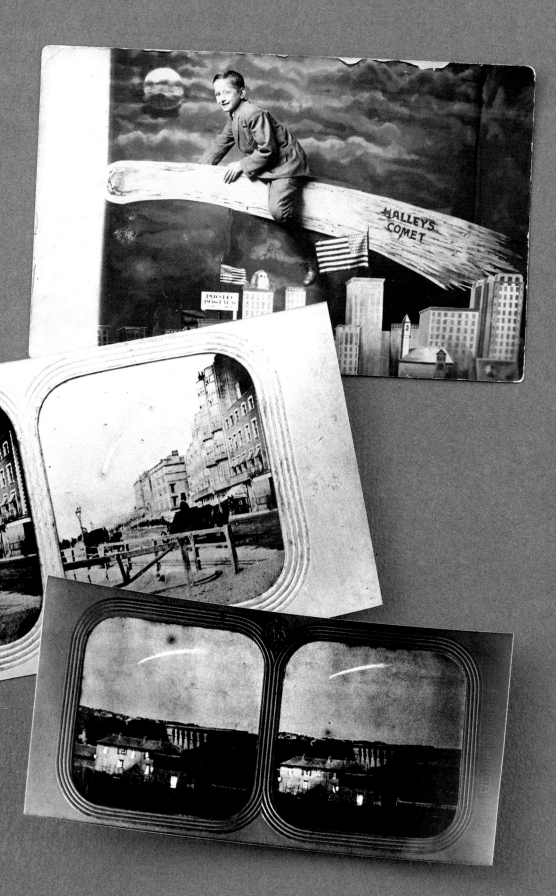

ELK MINSTRELS DEFY COMET

Superstition gets little consideration in Revere, at least with the members of the Revere Lodge of Elks. They proved it when they founded their lodge, "Baby Lodge," as it is called, on the 13th of the month. They proved it again recently when they planned to hold their first minstrel show on the two nights immediately following the day when the tail of Halley's comet is scheduled to whisk the earth. To be specific, the minstrel show of the Revere lodge, No. 1171, B. P. O. E., is to be held at Scenic Temple, Revere Beach, Thursday, May 19, and Friday, May 20. The committee has worked hard to make the first public appearance of the "Baby Lodge" a great success.

The committee includes David J. Hurley, Maffitt Flaherty, W. Pence Mitchell, G. Fred Nolan, F. S. Pfister and Herbert Watson. G. Fred Nolan is to be interlocutor. The soloists are David J. Hurley, F. S. Pfister and Fred Millwood.

Draper Is Not Alarmed By Comet or Democrats

Governor Draper said to-day:

"I have no more fear that the comet or its tail will do harm than I have that the Democratic party will be successful next Fall."

WOMAN SEES COMET'S SECOND VISIT

MIDDLEBORO, May 18.—Among the most enthusiastic comet watchers in this locality is Mrs. Eliza Reed of South Main street, who remembers distinctly the visit of Halley's comet seventy-five years ago. Mrs. Reed recently celebrated her eighty-ninth birthday and tells many interesting stories of early life in this section. She was fourteen years old when she saw the hazy visitor. She then lived at Provincetown, and on the hill where the Pilgrim monument now stands hundreds of visitors from all over the Cape watched the comet nightly.

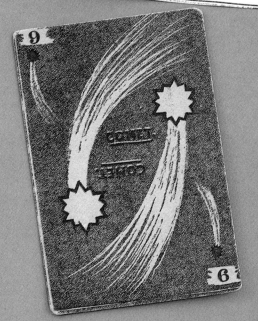

I. At whim of the wind in open air,
Rocking with gentle ease,
Cradled in a gossamer cloud,
Floats a star of happy promise.

LA COMÈTE

I. — Dans les airs au gré du vent   Bercée dans un fin nuage,
Se balance doucement   Une étoile au doux présage.

LA COMÈTE

II. — Vite prenez vos lorgnettes,   Qui là haut dans le grand ciel,
Savants pour voir la Comète   Se rit du pauvre mortel.

II. Scholars quickly seize your glasses,
If you wish to view the Comet,
Which way above the far horizon,
Casts its bright eye on poor mortals.

III. Bent beneath his weighty toil,
Our doughty laborer still smiles,
As he prepares a welcome harvest,
To grace this year's the Comet.

III. — Courbé sous le dur labeur,
Souris rude travailleur,
Bonne récolte s'apprête
C'est l'année de la Comète.

LA COMÈTE

IV. — L'Étoile éclairant les cieux
De ses rayons lumineux,
Nous apporte l'espérance
Des jours de paix, d'abondance.

LA COMÈTE

V. The Comet with its gentle fingers,
Brings good fortune to the vineyards,
And presents the vermilion-ed grape,
A nectar without equal.

This glorious star lighting the heavens,
With its luminous rays,
Fills us with reason to hope,
For peaceful and abundant days.

V. — Bonheur pour les vignerons,
La Comète aux doux rayons
Donne à la grappe vermeille,
Une saveur sans pareille.

LA COMÈTE

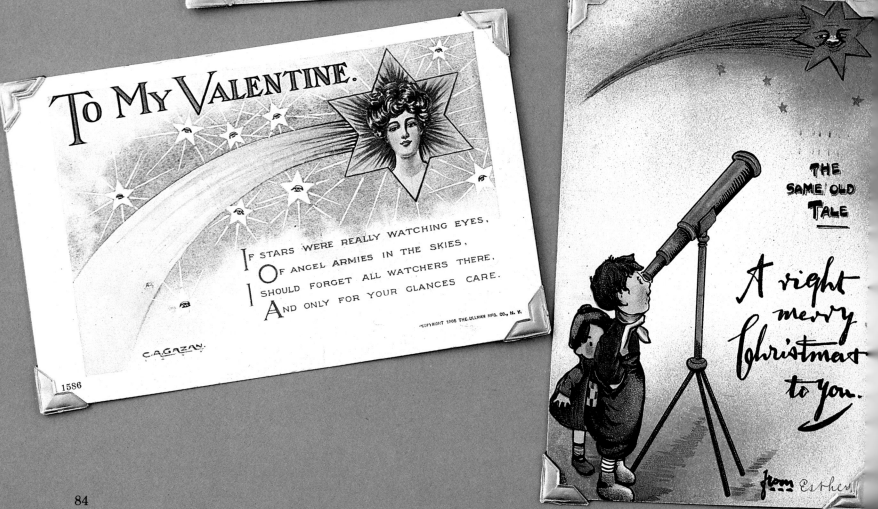

THURSDAY, MAY 19, 1910.

WELL, THE COMET'S TALE HAS PAST, AND EVERY ONE IS DOING BUSINESS

Continued From First Page.

made a celebration something in the nature of a Mardi Gras, on the East Side the occasion was taken more seriously.

Some of the more superstitious in the sections largely inhabited by foreigners were on the verge of panic, and the police were called upon at different times to calm them and disperse the gatherings. Many prayed on bended knees in the streets and parks, and several religious processions took place in different quarters of the city.

The crowds were denser on the East Side but Broadway and Riverside Drive held great crowds, many of them in autos and carriages, eager to witness any spectacle that the passing of the comet might afford.

The East Side crowds were overjoyed as time went on and nothing dreadful came to pass, while those on Riverside Drive in Central Park and on the roofs of the apartment houses and hotels finally left their posts keenly disappointed that the heavenly display predicted by some astronomers did not materialize.

Thousands of persons thronged the streets in the foreign quarters.

There are about 300,000 inhabitants in the lower East Side section, and it seemed as if every man, woman and child was in the streets. About 4,000 gathered in Mulberry Park, and many of them spent hours on their knees in prayer.

Suddenly one of the throng spied a large, bright light descending, and shrieked a warning to the others. A serious panic quickly threatened, and the police reserves had to be called out from the Mulberry street station to calm the crowd and clear the streets. They drove the frightened men and women into their houses, where they leaned from the windows, shouted and cried out while they watched the descent of the strange light, which proved to be a toy balloon.

Even this anti-climax to their fears did not serve to entirely quiet them, and hundreds spent the night in the streets.

The roofs of the New York hotels were crowded. No one, so far as could be learned, saw the comet, but there were lecturers to tell them all about it and the sky gazers had about as good a time as if they had actually seen it.

"Uncle Joe" Cannon led a party of fifty from the dinner of the National Association of Manufacturers to the roof, but they didn't stay long. There was nothing unusual to be seen, so they went back to the banquet hall.

A New and Becoming Style in
Silver Brand Collars
2 for 25c.

The Only Collars with Linocord Endless Buttonholes

Two Heights:
HALLEY, 2⅛ in.
COMET, 2⅜ in.
Quarter Sizes.

In collars the *Style, Fit and Comfort* depend on having strong, sound buttonholes. No matter how good the collar may be in every other particular, once a buttonhole is stretched or broken, you have a gaping, slovenly, ill-fitting collar.

THE LINOCORD BUTTONHOLE is made with an eyelet like the buttonhole in your coat, hence is pliable, easy to put on and off. It is *reinforced all around* with a *Stout Linen Cord*, which makes it so strong it can't possibly stretch or break, even with the hardest or longest wear—a SILVER Collar will always fit and look as the designer intended.

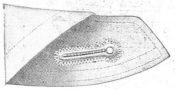

Linocord Endless Eyelet Buttonhole.

SEND FOR "WHAT'S WHAT."

The latest word in Men's Fashions. It embodies the dicta of the foremost fashion authorities with reference to every item of men's apparel. It not only tells what to wear but also what not to wear. Fully illustrated. Invaluable for reference. Yours for the asking.

GEO. P. IDE & CO., 498 River Street, Troy, New York.

BEVERLY

GRISWOLD

HARBORD

HARMON

THE QUIVER

CLEMAK — For Quick, Close Shaving

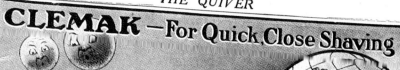

"*My word! that was a close shave*"

"Greased lightning isn't in it with the Clemak. Never saw anything like it in all my life. Just a dab or two with the lather brush, a pass or two with the Clemak, and I'm shaved. And best of it is, same thing happens every morning—no trouble ever."

CLEMAK Safety Razor 5/-

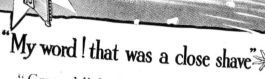

5/- Silver-plated Clemak Razor with stropping handle and seven blades.

Combination Outfit: A Triple Silver-plated Razor, Twelve specially selected Blades, Patent Stropping Machine, with Velvet Hide Strop

10/6

The Clemak is a sturdily-made British Safety Razor which combines unsurpassed efficiency with a simplicity quite unique. No razors shave better than the Clemak; none give less trouble. It's cleaned in a moment, stropped in a minute; there's nothing to remove, nothing to adjust.

"*NOW is the watchword of the wise*" —secure your Clemak TO-DAY.

OF ALL CUTLERS, STORES, &c.,
or from the
CLEMAK RAZOR CO., 17, Billiter Street, London, E.C.

The Clemak Book will interest you. Write for a Copy to-day.

W. PLANT & CO., Market St., Sydney, sole Australasian Agents.

THE GREATEST STAR EVER DISCOVERED.

Comet Pudding
Take a cold cup of milk,
    And dump it in your pot.
Put that on the stove,
    Til it's sure good and hot.
Add a little pinch of salt,
    And a touch of nutmeg.
Use a full cup of sugar,
    But just one brown egg.
Stir slowly and add,
    A smidgen of Comet.
One taste and your guests,
    Surely will...

If an advertiser could tie his product to the Comet's tail, everyone would see it as the Comet flew by.

LAMPE PHILIPS
On vient de constater que la comète de Halley n'est rien autre qu'une LAMPE PHILIPS
PHILIPS

"WHAT DRIVES THE COMET THROUGH THE SKIES!"
INQUIRING LITTLE JOHNNY CRIES.
"HOLD TO THE LIGHT MY SON, AND SEE
THE WONDERS OF ASTRONOMY."

THE PAINT KEGS OF THE WORLD ARE WAITING TO BE FILLED.
HARRISON'S "TOWN & COUNTRY" READY MIXED PAINTS
THE WORLD HAILS YOUR ADVENT WITH PLEASURE
GREETING
EUROPE
ASIA
AFRICA
HARRISON BROS & Co
"OH! SEE THE COMET."

BOSTON, FRIDAY, MAY 20, 1910.

GOING SOME!

COMET SPEEDWAY

COMET GOES ON VACATION FOR AWHILE

Halley's comet is now on vacation, the duration of which may be until to-morrow night or until Thursday evening. It was not observable to the thousands of Greater Boston people who climbed to their roofs this morning, and it will not be observable again until it has escaped the brilliancy of the sun, of which it is now a near neighbor. When it has shot across the face of the sun and has sped a few million miles on its other side it will regain its brilliancy. On Friday evening, after the last of the sun's rays have faded, the comet will be at its brightest, and astronomers predict a brilliant spectacle, contingent only upon clear skies.

The doubts as to the duration of the recess which the earth is experiencing are due to the possibility of happenings to-morrow night, when the tail of the big gaseous wanderer will be nearest the earth.

It cannot be said whether anyone on this side of the globe will know when it flashes by, for the scientific men who have cornered all the lore of the stars and planets and other celestial bodies don't agree as to the prospects.

Additional news of the comet appears on Page 3, Column 1.